Steck-Vaughn Shutterbug Books
MATH

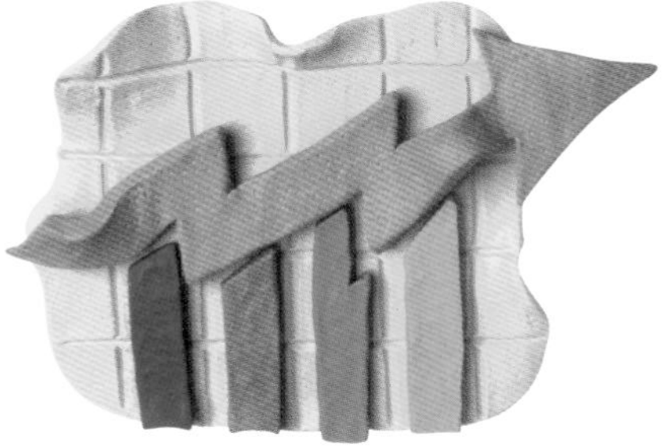

Charts and Graphs

by B. A. Shaver

STECK-VAUGHN
A Harcourt Company

www.steck-vaughn.com

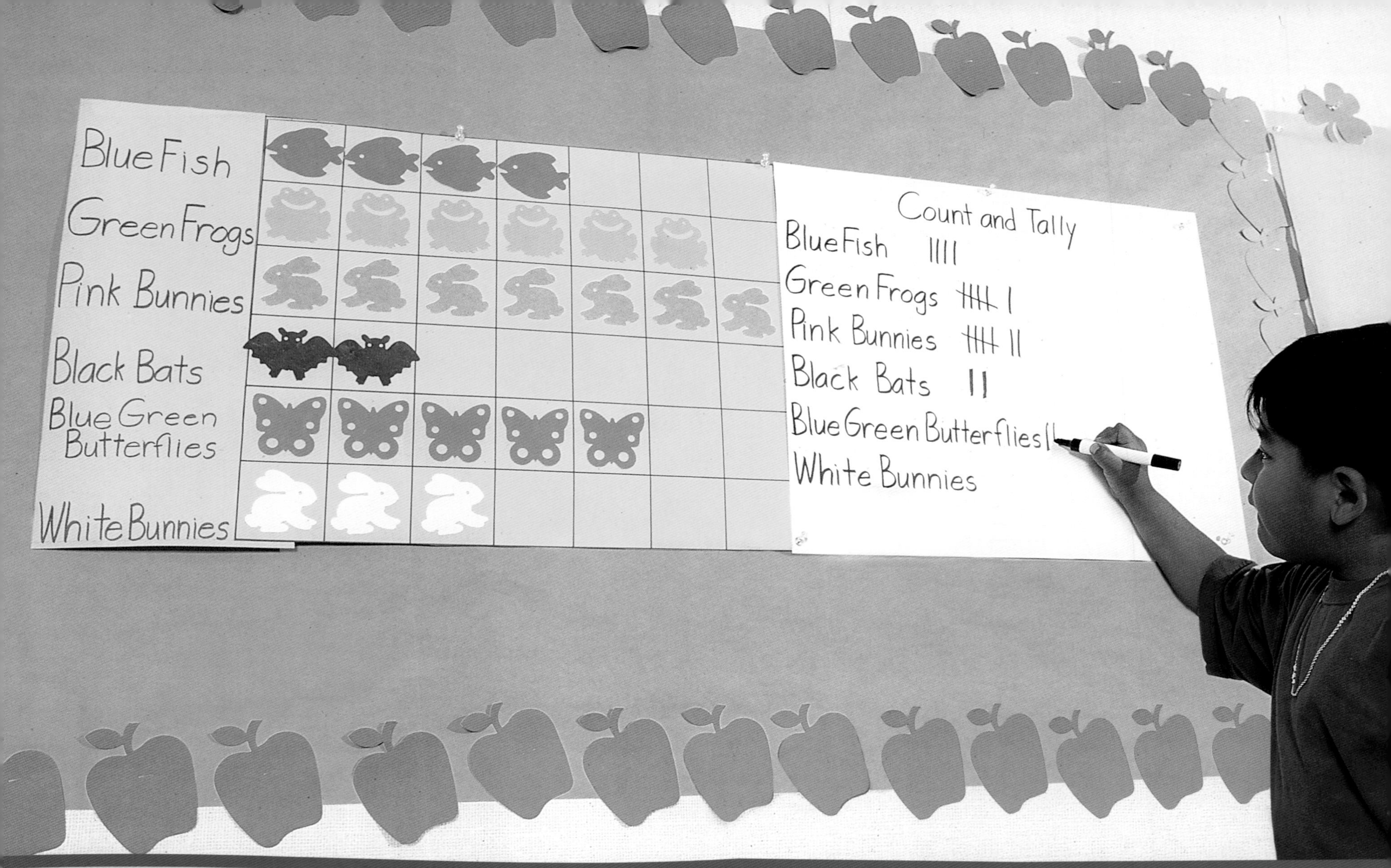
Blue Fish
Green Frogs
Pink Bunnies
Black Bats
Blue Green Butterflies
White Bunnies
Count and Tally
Blue Fish
Green Frogs
Pink Bunnies
Black Bats
Blue Green Butterflies
White Bunnies

Graphs and charts show the number of things.

Other kinds of graphs show the number of things, too.

Count the .
Count the .
Count the .

Colors of Marbles

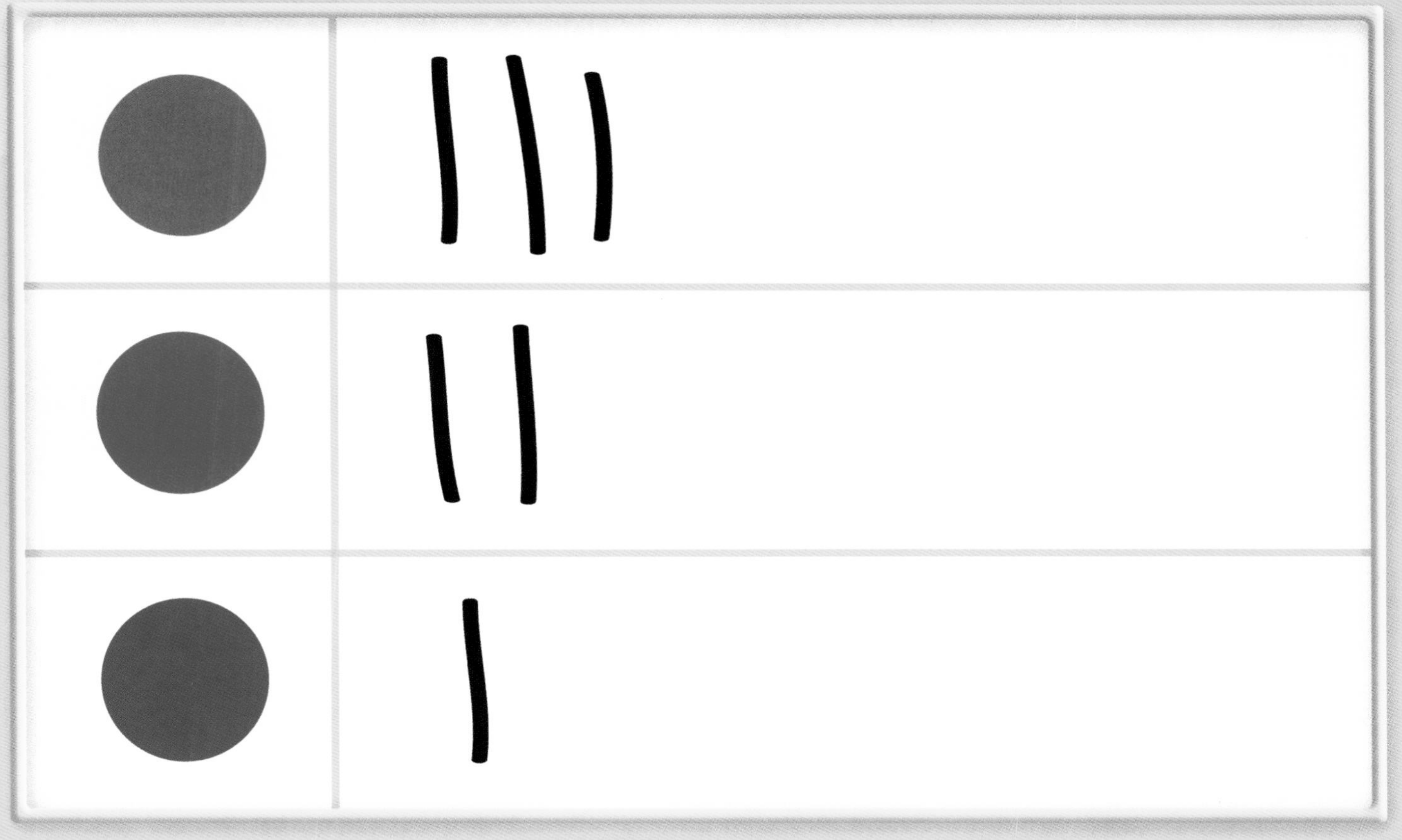

This chart shows the number of each color of marble.

Count the .
Count the .
Count the .

Kinds of Fruit

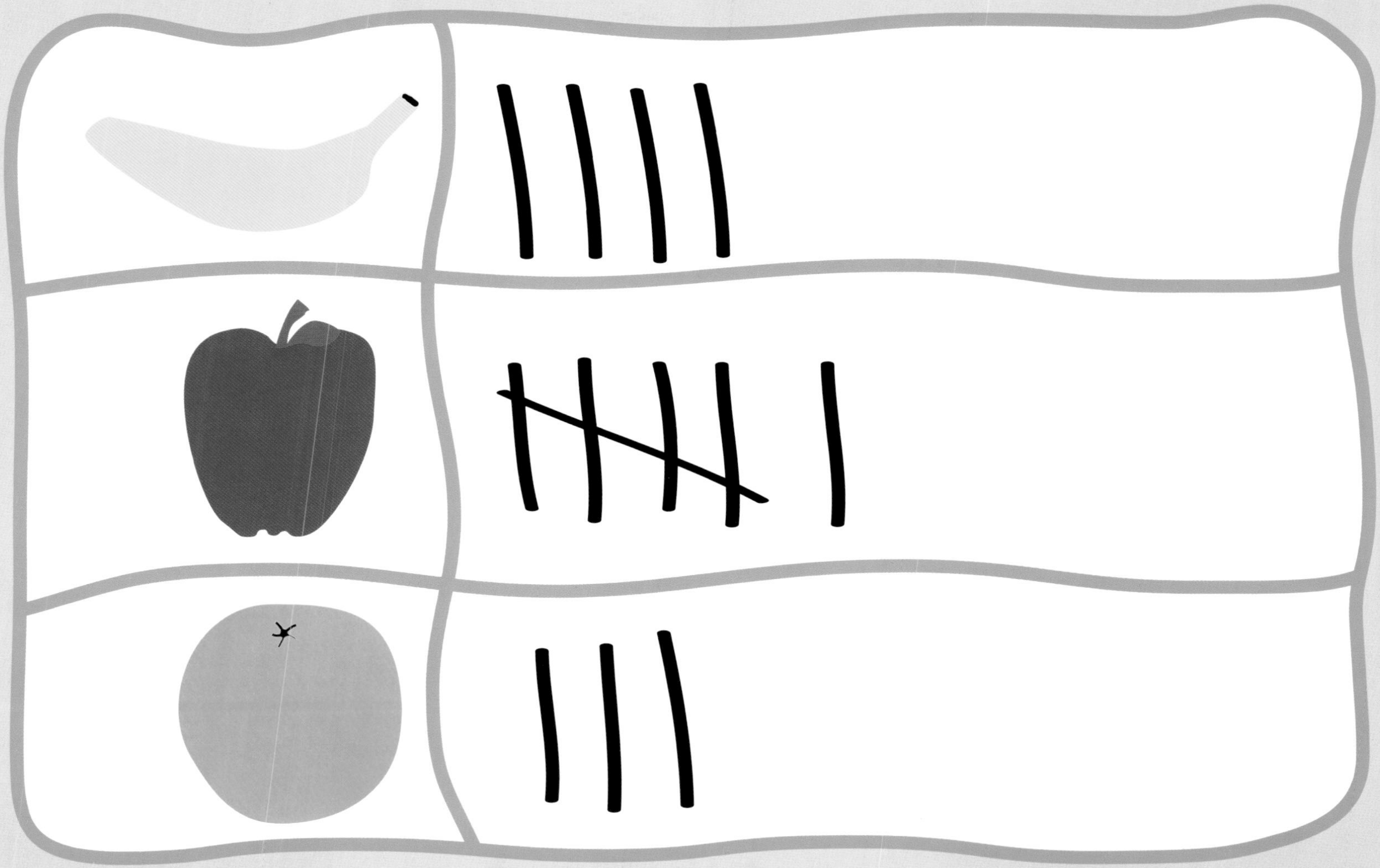

This chart shows the number of each kind of fruit.

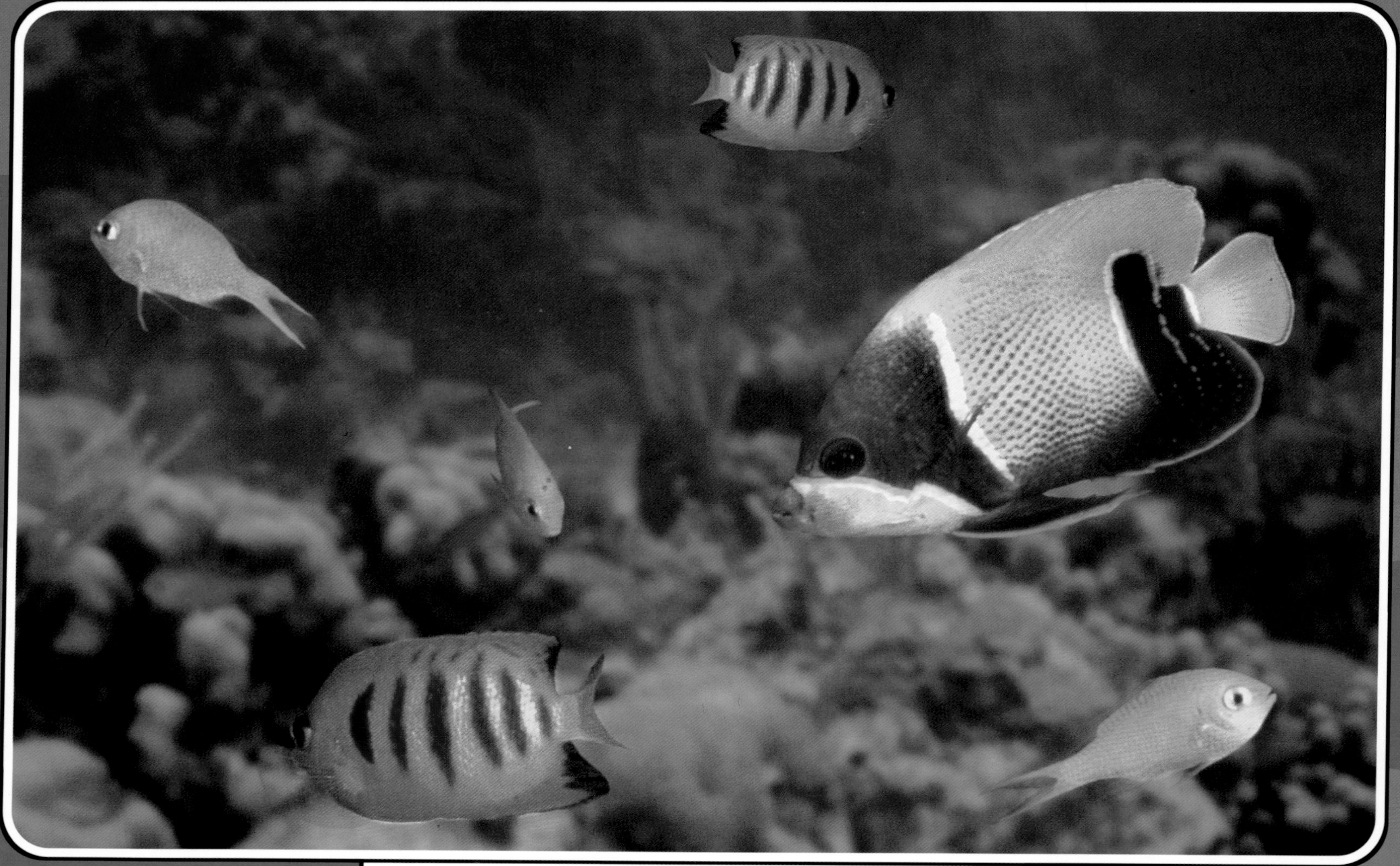

Count the  .
Count the  .
Count the  .

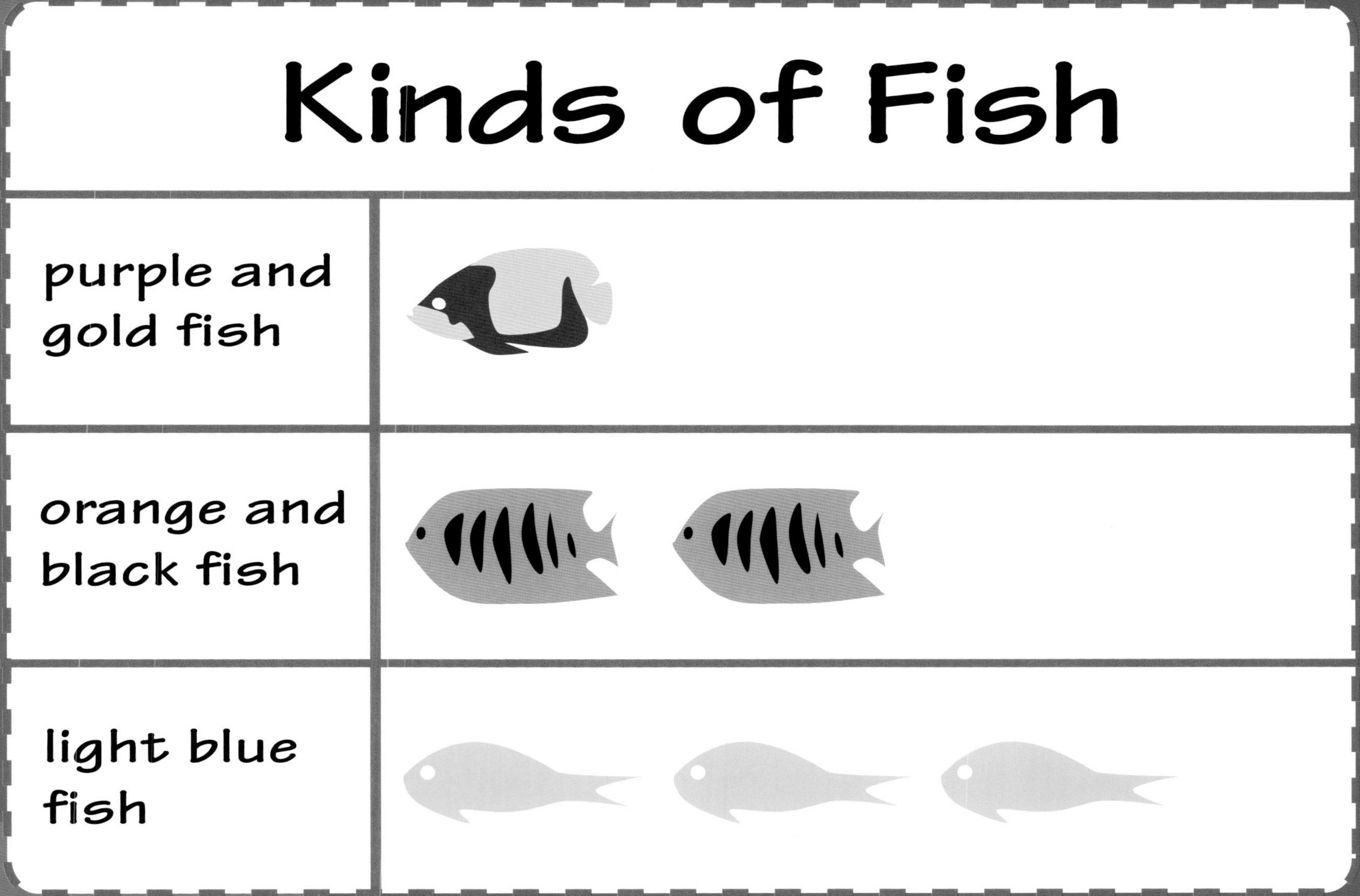
Kinds of Fish
purple and gold fish
orange and black fish
light blue fish

This graph shows the number of each kind of fish.

Count the .
Count the .
Count the .

Colors of Balloons
red
blue
orange

This graph shows the number of each color of balloon.

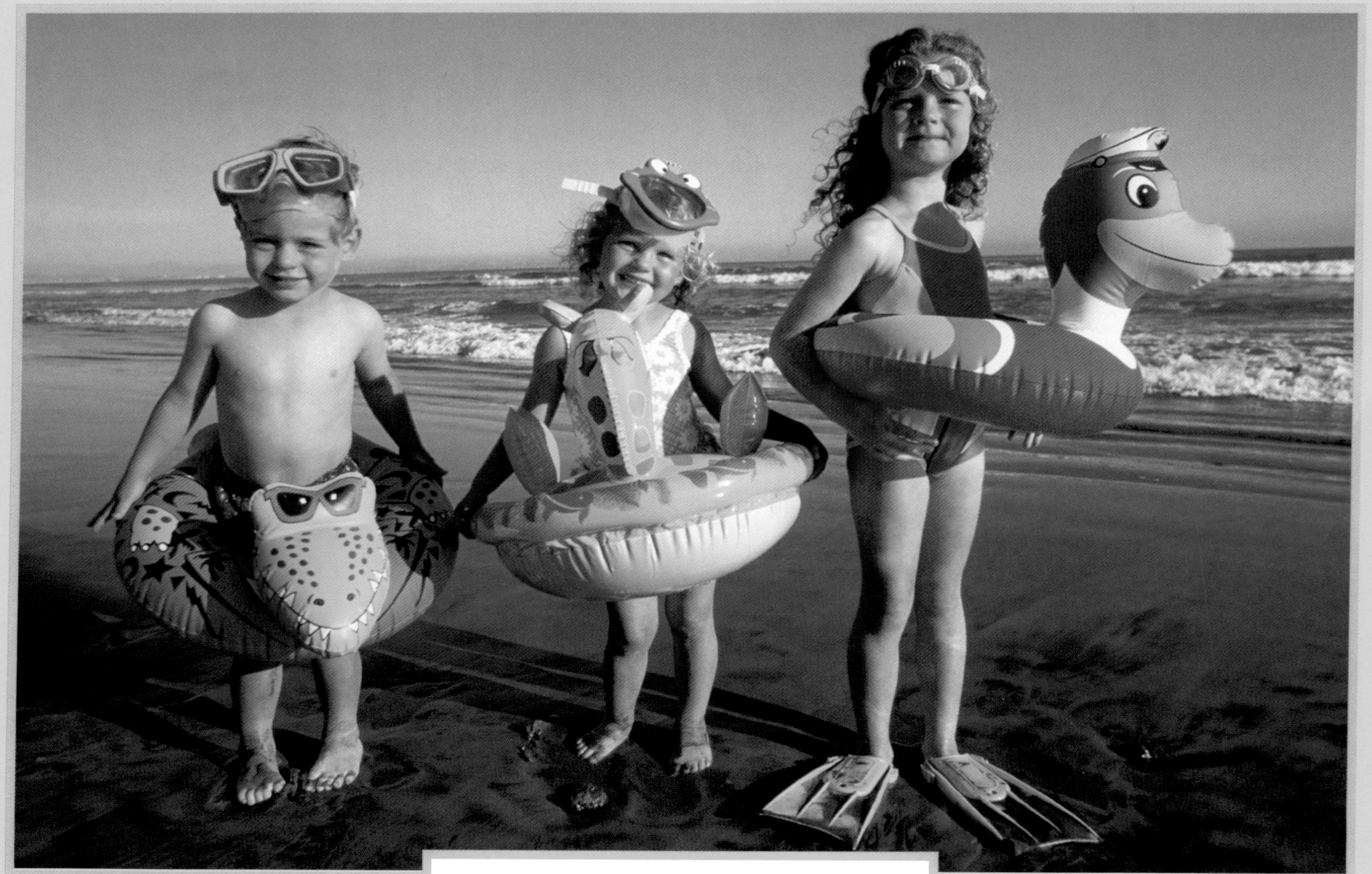

Count the boys.
Count the girls.

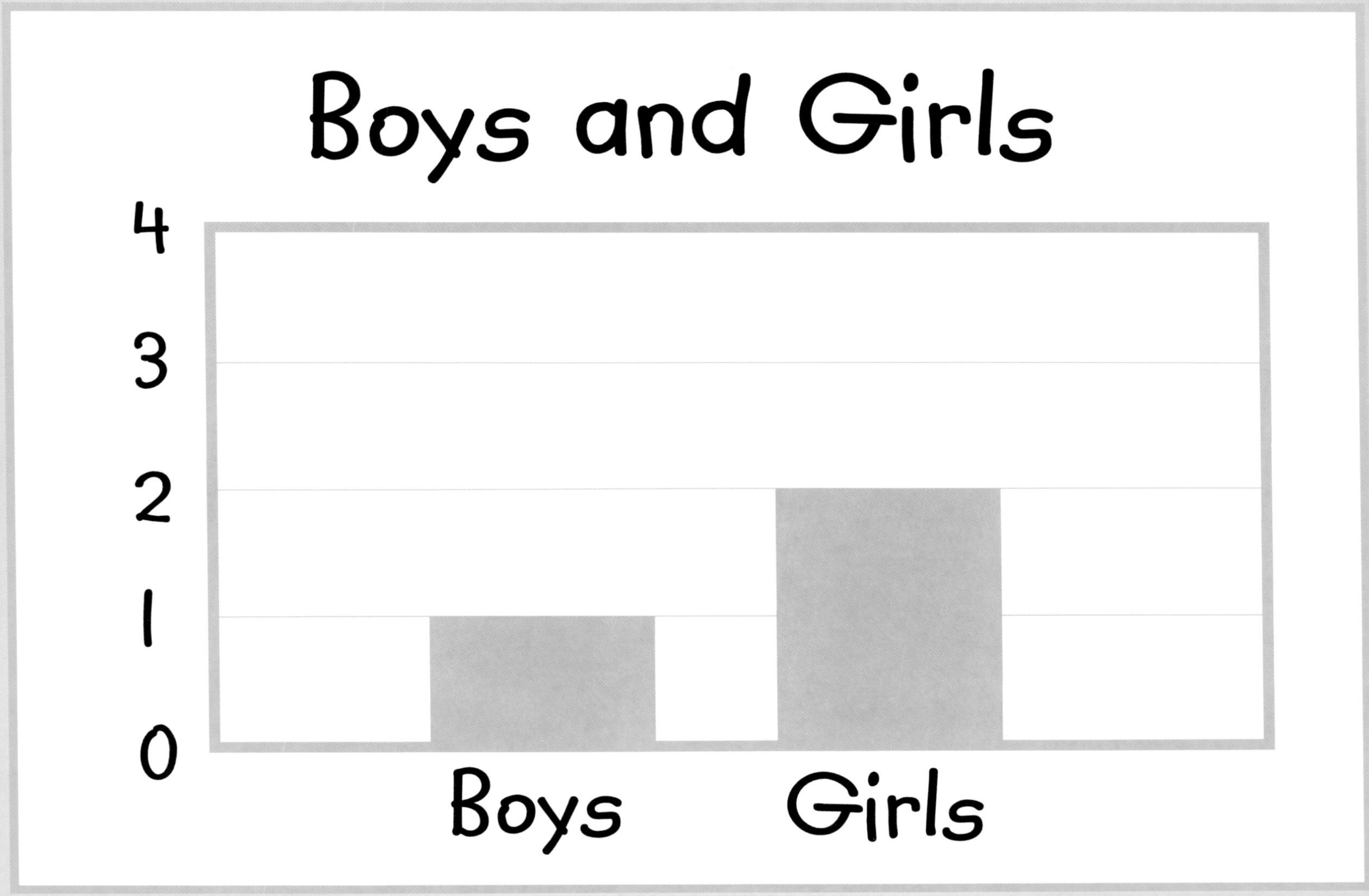

This graph shows the number of boys and girls.

Count the .
Count the .
Count the .
Count the .

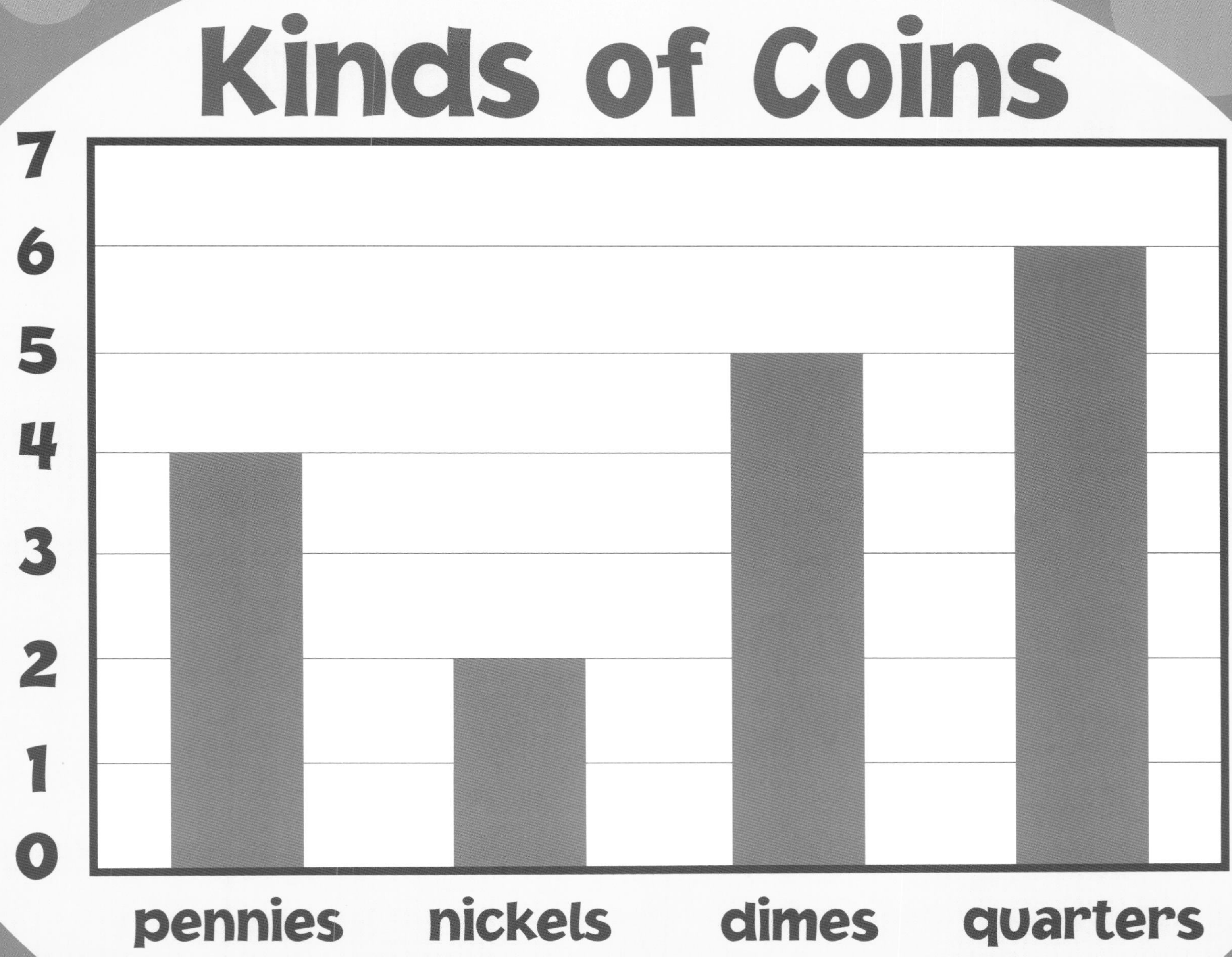

This graph shows the number of each coin.

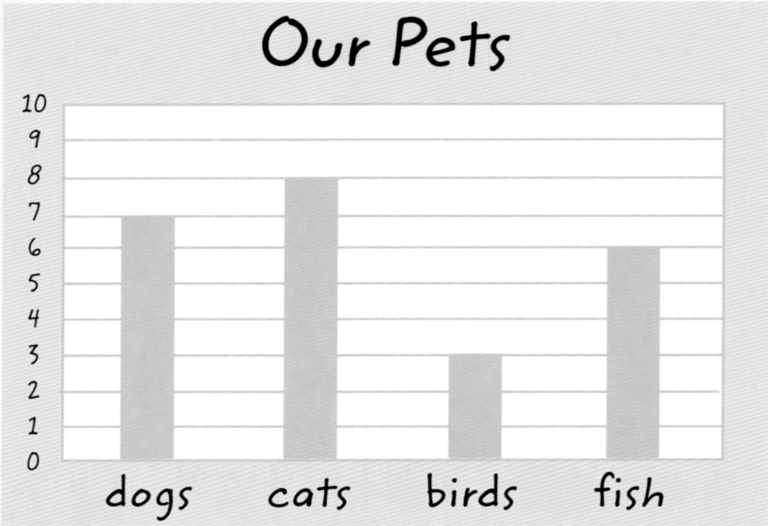

You can show the things you count in different ways.